submarine telecoms
FORUM

Offshore Energy Edition

90

WEBSITE TRAFFIC: UNIQUE VISITS

68,754
3-16

63,914
4-16

68,222
5-16

MAGAZINE DOWNLOADS

62,035

68,037

67,054

Issue #87 - Released 1-16

Issue #88 - Released 5-16

Issue #89 - Released 7-16

129,610
6-16

78,105
8-16

70,326
7-16

ALMANAC DOWNLOADS

Issue #18 - Released 5-16

498,058

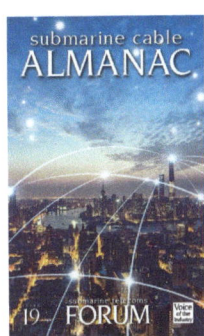

Issue #19 - Released 8-16

517,256

EXORDIUM

BY WAYNE NIELSEN

Welcome to Issue 90, our Offshore Energy edition.

If it's September, then what has been a whirl-wind implementation season should be coming to an end. But by all accounts, the industry is still going full bore, and projects that would normally be slowing down for the end of year, are still going strong.

I have had the chance to talk recently with a number of people in the industry – capacity guys, installers and manufacturers – and all seem to agree that the current pace of activities will continue into next year. And that, no doubt, is a good thing.

It's also surprising that with the offshore Oil & Gas market seemingly sliding into oblivion in the last year due to the price of a barrel,

the traditional commercial systems are still full speed ahead. But the Oil & Gas market has not been completely quiet. Nextgen has recently activated its $139 million North West Cable System in Australia, an ambitious 2,100km fiber optic submarine cable between Darwin and Port Headland with links to multiple offshore facilities. And more offshore systems in other regions are in the offing despite the price of oil.

So what does this mean – are energy companies being driven to submarine cable for reasons besides the barrel price?

The short answer may be yes.

We have known for some time that a major driver to offshore Telecoms is the need to reduce cost, and it seems that the low barrel price may be impacting future systems in two ways: One, systems have indeed been delayed, but not necessarily canceled; meaning plans are still afoot. Two, the relatively low barrel price is now driving forward Telecom systems that can help reduce production costs in order to stay competitive in the wider market for both existing and future facilities.

So, the dirge playing for the Oil & Gas submarine cable space may be a bit premature. We'll see.

Happy reading,

Wayne Nielsen is the Founder and Publisher of Submarine Telecoms Forum, and previously in 1991, founded and published "Soundings", a print magazine developed for then BT Marine. In 1998, he founded and published for SAIC the magazine, "Real Time", the industry's first electronic magazine. He has written a number of industry papers and articles over the years, and is the author of two published novels, <u>Semblance of Balance</u> (2002, 2014) and <u>Snake Dancer's Song</u> (2004).

 +1.703.444.2527

 wnielsen@subtelforum.com

IN THIS ISSUE...

ADVERTISER INDEX

News Now

➜ AAG Submarine Cable Damage Disrupts Internet Connectivity in Vietnam

➜ Bangladesh Submarine Cable Company's profit rises 28pc

➜ C&W Networks Selects Cologix to Unlock Traditional Connections Between North and South America

➜ CenturyLink Taps Aqua Comms for High-Capacity Connection Between New York and London

➜ Chile Publishes Bidding Terms for Patagonia Fibre Tenders

➜ Connectivity Slows as Vietnam Internet Cable Undergoes 10th Repair in Three Years

➜ Docomo Pacific Completes Marine Surveys for ATISA Deployment

➜ French Guyana Invites Tenders for Subsea Cable

News Now

→ Internet in Vietnam to Resume in at least 10 Days

→ MENA Upgrades Subsea Network With Infinera Connecting Mediterranean and Middle East

→ Mobily Launches Submarine Cable AAE-1

→ Monet Submarine Cable Selects Equinix for Landing Station and Next-Generation Digital Gateway to Latin America

→ Nexans Supplies Cable to Telenor Norway for Lofoten Link

→ Nextgen Switches on North West Subsea Cable System

→ Ooredoo Launches AAE-1 Submarine Cable in Qatar

→ Proposed FCC Rules for Team Telecom Review of Applications with Foreign Ownership, Shorten Timeframes, Add Burdens

- PT.Telkom Selects NEC to Build the "Indonesia Global Gateway (IGG)" Submarine Cable

- Regulator Calls for Lower Telefonica Wholesale Tariffs

- Reliance Jio Gets Green Nod for AAE-I Subsea Cable Project

- Seaborn Networks Selects Amdocs Optima to Monetize its Submarine Cable Infrastructure

- SKR1M Faster Internet Services for Sabah, Sarawak by June 2017

- SLT Gets Full Landing Status for SEA-ME-WE 5 in Matara

- Small Polynesian Nations Look to Submarine Cable Link

- Submarine Cables and BBNJ: ICPC Publishes White Paper and Presents to the United Nations

News Now

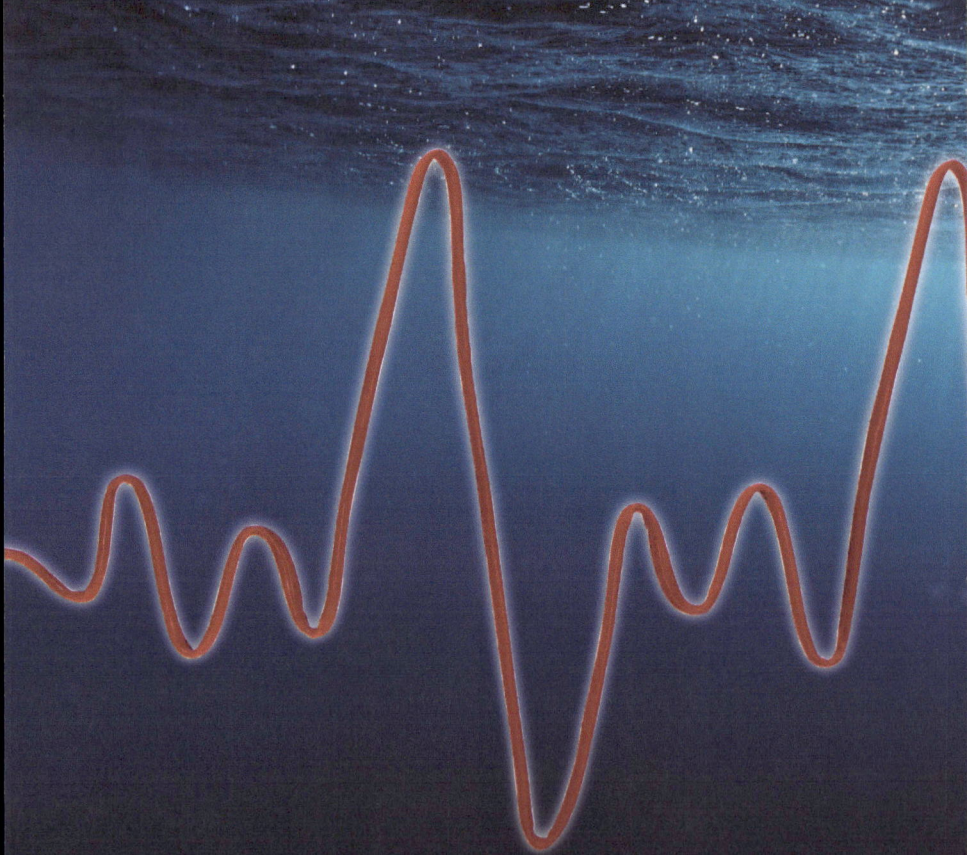

SUBMARINE CABLE SYSTEM REPORTING

Presenting the industry's most extensive collection of 375+ current and planned submarine cable systems impacting financiers, carriers, cable owners, system suppliers, component manufacturers and marine contractors, and detailing more than 50 menu-based data fields and maps in a customer-customizable report.

REPORTING IN 5 BUSINESS DAYS

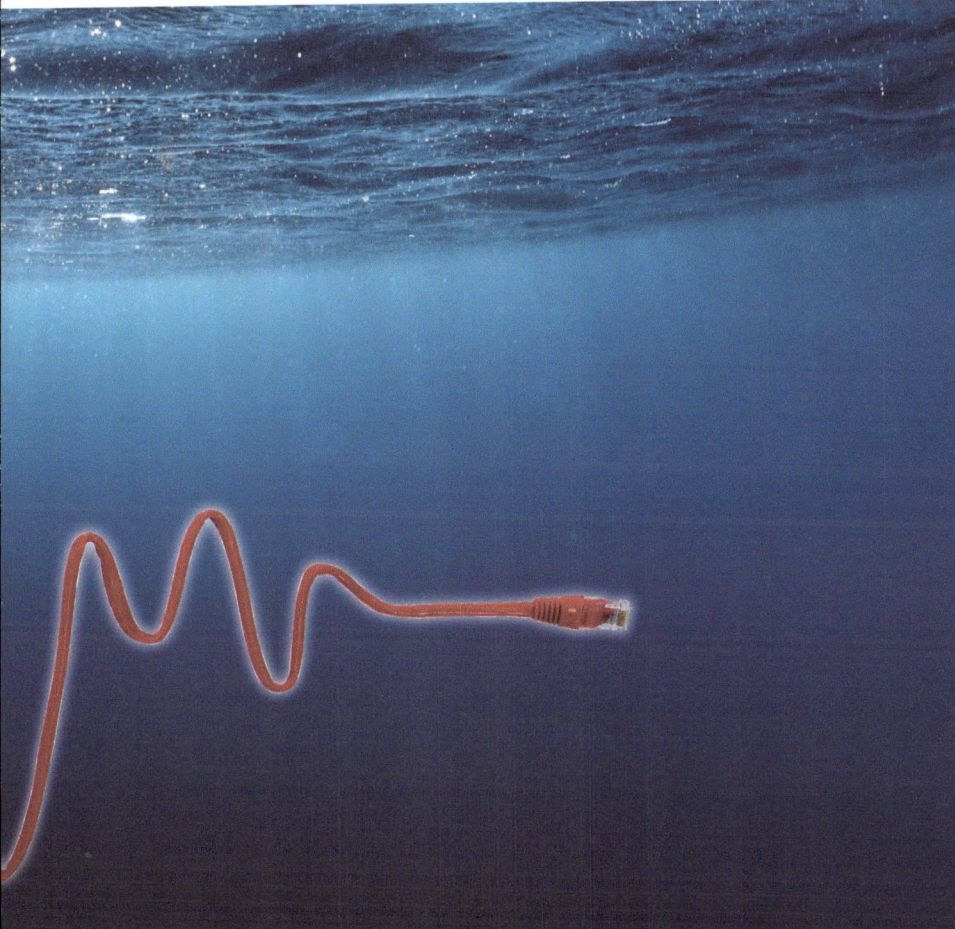

OFFSHORE ENERGY OUTLOOK

BY KIERAN CLARK

Since last year's Offshore Energy issue, there has only been a 10 percent decrease in planned systems for 2016-2019. This is a significant improvement, compared to last year's 61 percent drop in planned systems. Oil prices have seemingly stabilized, and are potentially improving — a sign of recovery for the offshore energy industry. Overall, the industry is still working to recover for the massive depression of oil prices since their last peak in 2014.

Welcome to SubTel Forum's annual Offshore Energy issue. This month, we'll take a look at the market for submarine fiber in the world of offshore energy platforms. The data used in this article is obtained from the public domain and is tracked by the ever evolving STF Analytics database, where products like the Almanac, Cable Map, Online Cable Map and Industry Report find their roots.

For last year's Offshore Energy issue, 5 systems were planned to be ready for service here in 2016. There were 2 more systems were planned for 2017, 8 systems planned for 2018

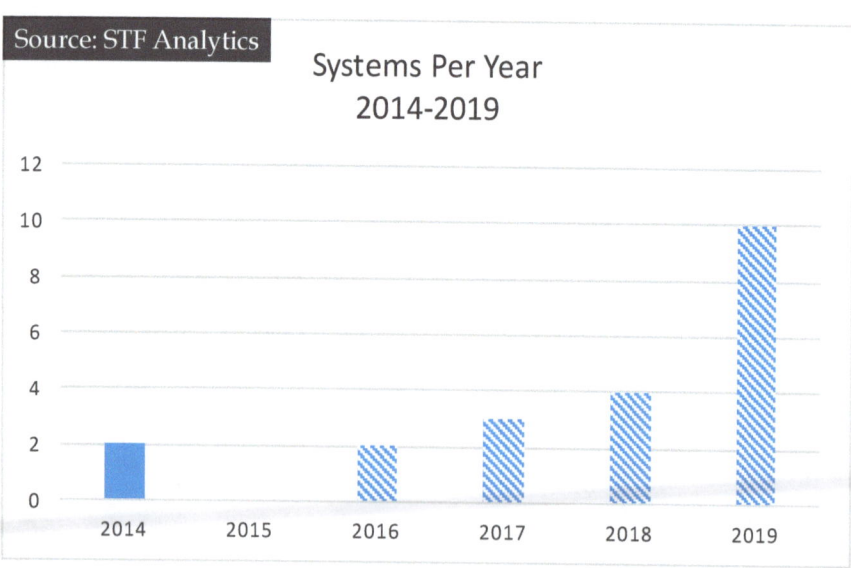

Source: STF Analytics

Systems Per Year
2014-2019

and 5 planned for 2019. After a year's time, these numbers have shuffled considerably. The change in planned systems count has resulted in 2 systems being implemented this year, 3 systems planned for 2017, and 4 systems planned for 2018.

As a result of some systems being delayed compared to this time last year, 2018 has dropped to 4 planned systems and 2019 can now expect 10 new systems. With new systems ultimately being tied to the price of oil, all of these numbers are subject to change based on the whim of the markets.

If oil prices remain low, expect more systems to be delayed or die out. However, if the recent price increases indicate a new trend system builds can be expected to increase.

With a cautious outlook for the next 4 years, the length of cable added annually follows a similar trend compared to last year's data, plus a massive uptick in 2019. A 2,500-kilometer spike is observed in 2016 thanks to a single 2,000-kilometer system in the Austral-Asia region while 2017 and 2018 show more moderate growth. An addition of 700 kilometers

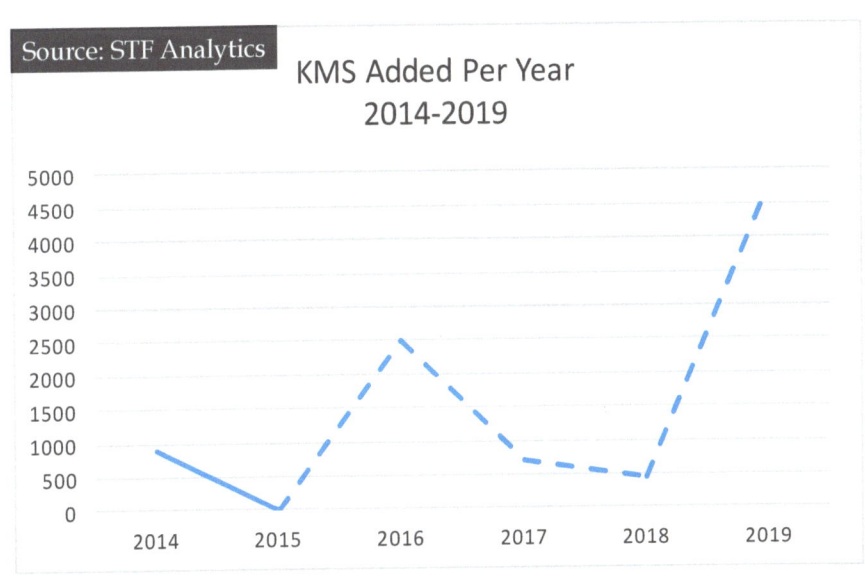

Source: STF Analytics

KMS Added Per Year
2014-2019

Source:
STF Analytics

Systems Announced Per Region
2016-2019

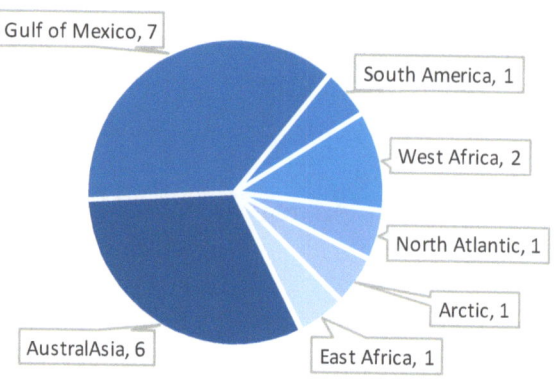

- Gulf of Mexico, 7
- South America, 1
- West Africa, 2
- North Atlantic, 1
- Arctic, 1
- East Africa, 1
- AustralAsia, 6

is projected for 2017, 400 kilometers for 2018, and a large jump of over 4,700 kilometers for 2019. If the price of oil continues recent trends, expect these numbers to remain steady or slightly increase.

From a regional perspective, the Gulf of Mexico and Austral-Asia regions will be the busiest over the next several years. The two regions combined are set to account for 67 percent of all planned systems for 2016-2019. West Africa will see

a total of 2 new systems, with most of the remaining regions of the world seeing only a single new system in their future. These numbers closely parallel those of a year ago, and follow trends in the general oil industry.

While the offshore energy industry at large has slowed down in recent years, the Gulf of Mexico and Austral-Asia regions continue to see high levels of expansion. It is this growth that drives these regions to the top of the pile with regards to new system activity, and should continue to do so over the next few years.

This time last year, the total estimated cost of systems planned for 2016-2019 was $1.045 billion. One year later that number has only slightly increased to just over $1.1 billion. Over $800 billion of this planned investment is in 2019 alone. This has largely been the result of multiple system delays, especially to the large and expensive systems in Africa.

As expected from the number of systems planned per region, the Gulf of Mexico and Austral-Asia regions make up significant portions of that $1.1 billion investment. Combined, they account for over $450 million.

 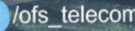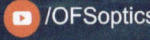

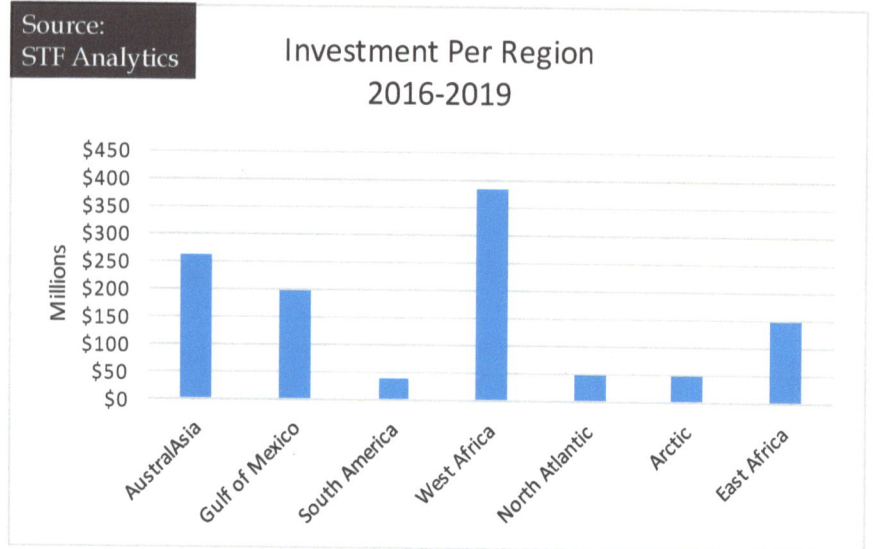

Investment Per Region
2016-2019

Millions

$450
$400
$350
$300
$250
$200
$150
$100
$50
$0

AustralAsia Gulf of Mexico South America West Africa North Atlantic Arctic East Africa

The West African region — which will see only 2 new systems over the next 4 years — will end up accounting for more than one-third of total system investment at a value of $383 million. While the number of proposed systems in the region is low, West Africa is expected to see 3,500 kilometers of cable added in 2019. This number accounts for over 40 percent of all planned cable and is the primary reason for such a large dollar investment in the region.

Most of the other regions that only have a single system planned naturally have much lower projections, combining for a total of $287 million. Notably, the Mediterranean and Black Sea regions show no activity planned for the next several years.

Dedicated systems are those built primarily by 1 or more Oil & Gas companies to serve their specific offshore facility's needs. Managed systems are those built by a Telecom service provider to 1 or more Oil & Gas companies' offshore facilities.

A year ago, 79% of systems planned for 2016-2018 were dedicated, with 21%

submarine telecoms
FORUM
UPDATE YOUR CONTACT INFO

http://bit.ly/29DBZrN

being managed. This year, that split has continued nearly unchanged. A full 75 percent of all planned systems for 2016-2018 will be dedicated. This trend should stay constant through 2019, with 71 percent of planned systems for the next 4 years being dedicated.

As companies push further out and explore new areas for drilling, they can rely less and less on existing systems managed by telecom providers. With most of the heavy growth in offshore energy happening in previously untapped areas, expect the prevalence of dedicated systems to continue.

While oil prices are still significantly below those of the last 2014 peak, they have recovered significantly from a year ago. The market has seemingly adjusted to this new normal, with prices having remained within the $40-$50 range for some time now.

Unfortunately, there does seem to be some concern over another supply glut within the next year or so as several countries have indicated a desire to increase production. Additionally, OPEC members have

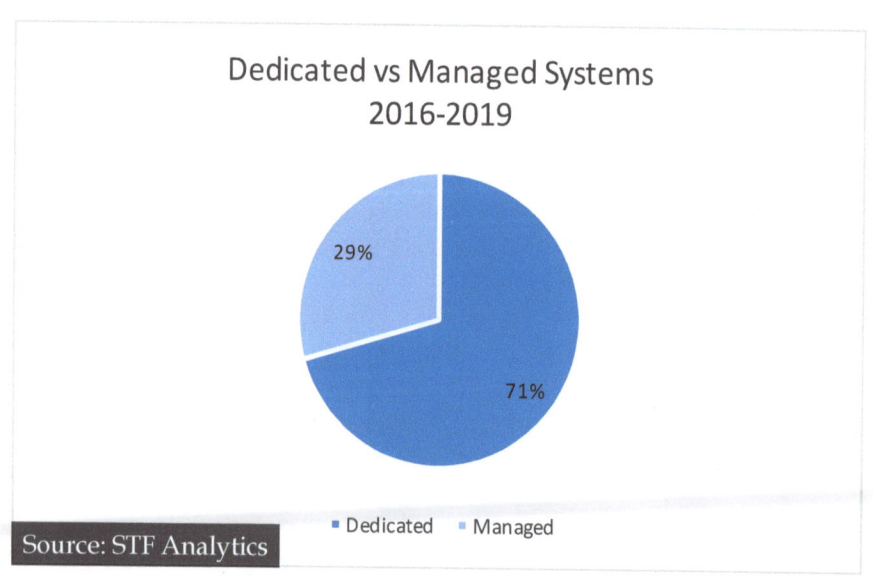

Dedicated vs Managed Systems 2016-2019

29%

71%

Dedicated Managed

been unable to agree on production caps. On top of all this, weak global economies have led to a further reduction in energy demand. There seems to be little indication that the supply erosion market analysts have insisted is coming will start any time soon.

Ultimately, outlook for this aspect of the submarine fiber industry remains almost entirely unchanged from a year ago. The rapid decline of oil prices from two years ago is still affecting the industry at large. There has been very little change in total planned investment, however more of that money has been pushed to later timeframes than initially expected. This part of the submarine fiber industry lives and dies by the price of oil, and as long as it remains at this new normal the growth will continue to be muted.

Kieran Clark is an Analyst for Submarine Telecoms Forum. He joined the company in 2013 as a Broadcast Technician to provide support for live event video streaming. In 2014, Kieran was promoted to Analyst and is currently responsible for the research and maintenance that supports the SubTel Forum International Submarine Cable Database; his analysis is featured in almost the entire array of SubTel Forum publications. He has 4+ years of live production experience and has worked alongside some of the premier organizations in video web streaming.

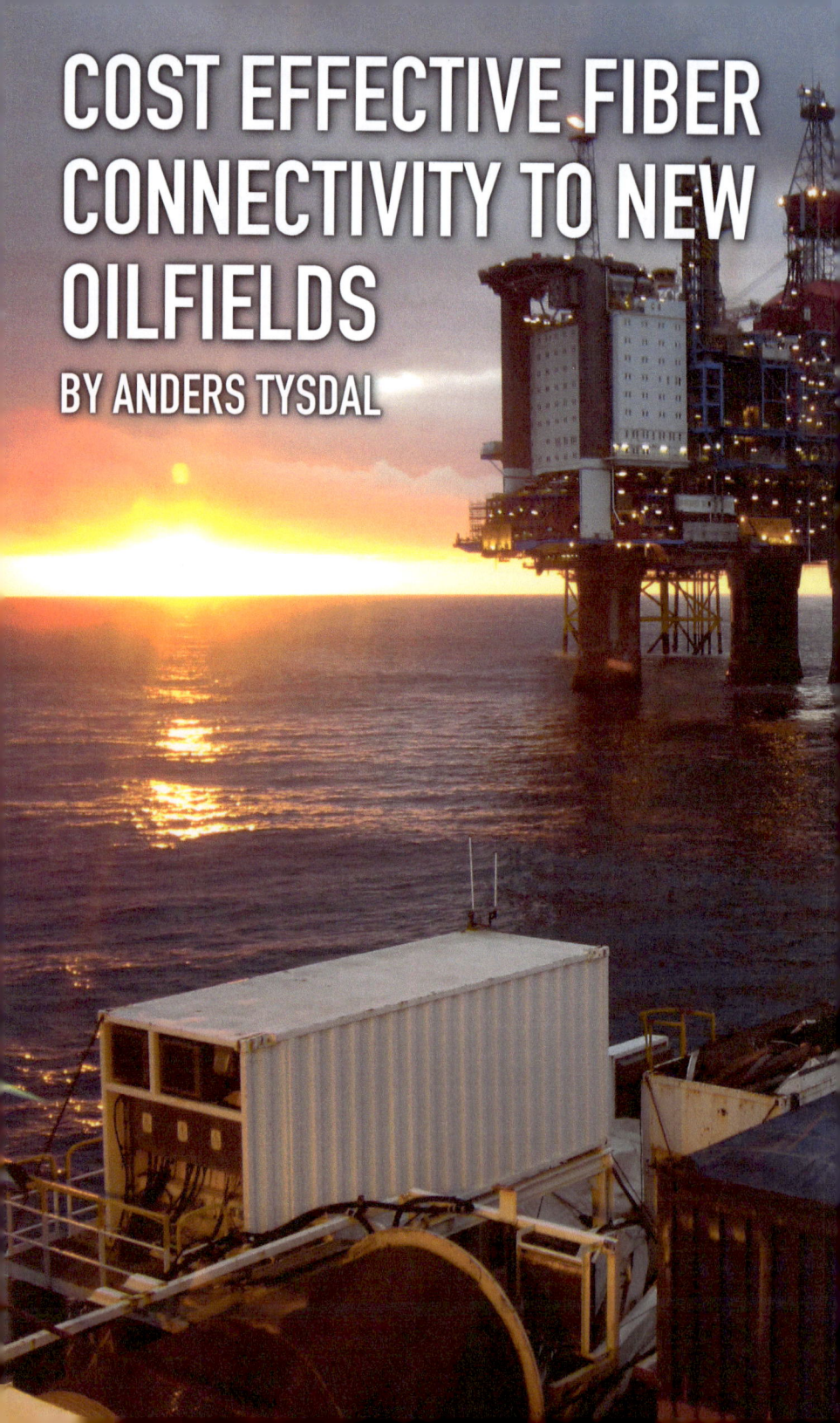

COST EFFECTIVE FIBER CONNECTIVITY TO NEW OILFIELDS

BY ANDERS TYSDAL

The exploration for offshore energy resources is becoming ever more challenging, as energy companies are forced to seek out new deposits in deeper waters ever further from shore. With more complicated exploration and operations, coupled with pressures on oil price, the oil companies are seeking new and smarter ways of operating through the use of new technologies.

Fiber optic networks, and the digital ecosystem that they enable and interconnect, are facilitating this enormous challenge by allowing corporations to more efficiently manage their offshore assets such as production platforms, subsea well management and drilling operations. In-depth analysis and understanding of oceanic, geological, and atmospheric data provides necessary insights for better decision making and utilization of IT resources, equipment and human capital. The transportation of this data requires reliable networks with high capacity and low-latency. Offshore energy corporations can leverage submarine networks to connect offshore assets to onshore assets to realize a variety of operational benefits to help differentiate and gain higher operational efficiency.

Most operators of larger, new field developments, particularly those in the North Sea, Gulf of Mexico, North-Western Australia and Brazil, regard fiber connectivity to the beach as a fundamental necessity for the operation of a modern

oil field. However, fiber to a new oilfield poses several challenges.

The cost of laying subsea fiber for a single field development to shore can often be prohibitive, particularly as distances in excess of 200km are not uncommon. The issue of redundancy is also a big concern. Basing ones' operations on subsea fiber is one thing, but relying on satellite connectivity as backup to such a way of working is risky. The high latency and low bandwidth available on satellite means operations will be seriously impaired in a backup sit-uation. The obvious way of overcoming the backup challenges would be to create a fiber ring system, or loop, for the development, which in turn would almost double the cost of an already expensive fiber installation project.

Adding a fiber project to an already stretched project team poses several challenges and adds to the overall project risk. In addition to the actual fiber installation project, the operator will have to deal with multiple legal and regulatory issues, beach crossing arrangements, pipeline and cable crossing agreements, environmental permits, decommissioning liabilities and so forth.

Management and maintenance of the fiber optic infrastructure poses challenges for an oil company; typically, operating a subsea fiber optic network is not core business at any oil company, so how does one ensure the availability of qualified telecom personnel and expertise, how are potential fiber breaks handled – and how does an oil company justify these costs?

The Tampnet Infrastructure

Tampnet, established in 2001, currently operates the largest offshore multi-terabit, low latency subsea optical fiber network in the North Sea and the Gulf of Mexico, which reliably serves over 240 offshore assets such mobile rigs, Oil&Gas FPSO (Floating Production, Storage, and Offloading) platforms, and exploration rigs. Reliable, high-speed, low latency network services are the primary goals of their network, which includes 2,500km of submarine fiber optic links, multiple strategically located 4G LTE base stations, as well as nearly 100 traditional point-to-point radio links.

One of the unique benefits of Tampnet's network from a reliability viewpoint is that their subsea North Sea network is unrepeated, and only contains passive fiber

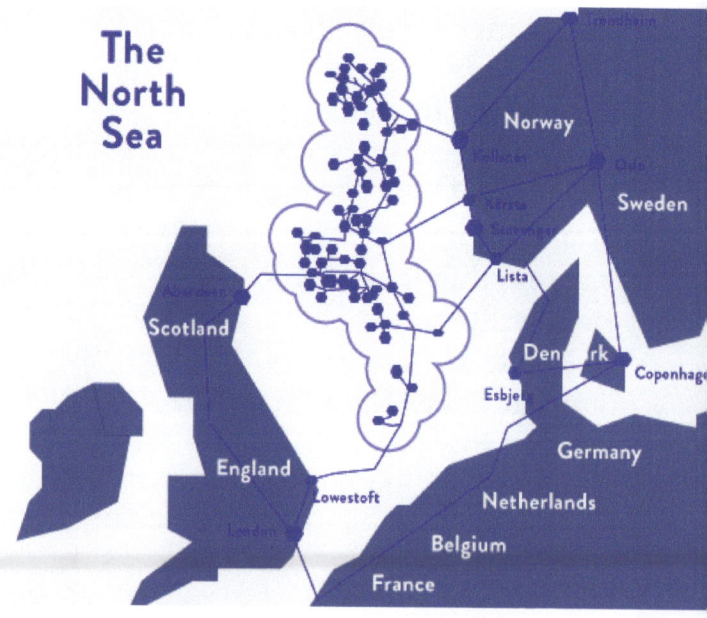

strands and no powered electronics. Although modern submarine network repeaters are quite reliable, they are less so than strictly passive (unpowered) fiber-optic cables. Unrepeatered networks do not require onshore Power Feed Equipment (PFE).. In short, the fewer active units in a network, the less things can go wrong, which means a more reliable overall network design is achieved. Also, full flexibility in technology developments is possible, by simply replacing topside, dry, terminal equipment. Customers are able to enjoy the inherent reliability benefits associated with passive submarine networks because Tampnet has deployed the very latest in Submarine Line Terminating Equipment (SLTE) that offers significant unrepeated reaches within and across the North Sea.

The network has been designed with future expansions in mind by installing a number of branching units/subsea tails in strategic locations.

The Tampnet Model for Connectivity to New Oil Field Developments

The Tampnet model for connectivity of new offshore oilfields attempts to address and overcome all of the main challenges of offshore fiber connectivity. Over the last 24 months Tampnet has successfully connected 5 new oil fields to its fiber infrastructure, while another two are currently in the planning. All of these projects have been completed within the proj-

ect schedule, within budget and with zero serious HSSE incidents.

The existing branching units placed on the seabed enable fiber connectivity with much shorter fiber routes than by going to shore. Full redundancy over divergent routes has been delivered – either on fiber or through one radio link hop to a secondary fiber path. Whereas most of these fields would have to undertake projects with cable lengths in the region of 250-300km for redundant fiber connectivity each, Tampnet has been able to offer the same with cable installed distances from 35-100km per asset.

The fiber projects are offered as turnkey projects whereby Tampnet takes full responsibility for the actual fiber project from its branching unit to the topside telecom equipment room. We get involved as early as possible in the engineering phase, in order to optimize the methodology for connecting the fiber topside. By getting involved with subsea engineering resources and the project team, other subsea structures and subsea activities near the Oil/Gas asset can be taken into account to optimize the way of connecting well in advance of the actual fiber project.

The overall cost for the project is significantly reduced as the length of new fiber required is normally reduced by 80% or more. Also – further synergies can be reaped as new field developments in the same region can be tied into the same fiber project. The perfect example of this is the connection of the Gina Krog, Edvard Grieg and Ivar Aasen fields in the North Sea. Three different field developments operated by three different oil companies connected in a single fiber installation campaign.

These fiber projects represent an expansion of Tampnet's overall infrastructure – which in turn may open up future opportunities.

Additionally, the new fiber combined with the new platforms represent additional sites for the installation of 4G LTE base stations – serving all mobile activity in the area, such as vessels, rigs, handheld devices and personal mobile devices. For this reason, Tampnet will typically also carry a significant portion of the capex cost of the project.

Managing crossing agreements in a basin can be an intricate and time consuming affair. Tampnet has the experience and expertise for this and takes on these responsibilities as part of the turnkey project. Other regulatory approvals, beach landing approvals are also a natural part of the Tampnet scope.

Tampnet uses highly specialized telecom cable laying vessels for the extension of their fiber network. Not only are the vessels more specialized, but often also comes at a more attractive price point than that of some typical offshore work vessels designed for oil &

gas subsea work. The pull-in of the fiber to the new offshore platform, or the subsea tie-in, will also be handled by the cable vessel, or by a vessel working for the Oil/Gas company at a later date.

Once the platforms is in situ and the topside fiber connection completed, Tampnet installs all necessary telecommunications equipment in the telecommunications equipment rooms. If the platform offers dual/ redundant equipment rooms, Tampnet will install its equipment in both locations. All telecommunications equipment will be owned and operated by Tampnet and made ready for provision of services for

the oil company in question – typically in the form of a redundant Private Ethernet service.

Redundancy is provided in the form of dual fiber connectivity or through a microwave radio link pointing at a separate platform with fiber connectivity. Albeit microwave radio links have their limitations in relation to fiber, they offer a very robust and reliable backup option. When installed with space diversity and operating on dual frequencies, the radio link backup offers bandwidth in the region of 300-500Mb/s and with availability in the region of 99.99% or higher.

The Tampnet Support Model

All telecom equipment is enrolled in the overall portfolio and therefore managed and monitored 24/7/365 by our highly offshore specialized Network Operations Centers. The personnel manning the Tampnet NOC are highly qualified telecom engineers, often with hands-on offshore telecom experience.

Although usually well protected through burial, there is always a certain amount of risk that over time the cable will become exposed on the seabed. This in turn means there is a risk of fiber cable breaks as a result of fishing activity or anchor drag. Locating, repairing and re-burying subsea fiber cable is a complex and costly operation that requires specialized vessels and skilled people on board. 24/7 Availability of such vessels is also an issue. For Tampnet customers this is not a concern. In addition to our focus on redundancy, Tampnet is also a member of the Atlantic Cable Maintenance Agreement (ACMA), providing instant access to capable repair ships on a continuous basis.

Tampnet's commercial model and support model is based on the fact that operators of the new oil fields are not in need of fiber cables – they are in need of reliable, low-latency and high-capacity connectivity services to their assets. Furthermore, they are interested in anything that can reduce cost and remove risk and uncertainties with regards to their future data connectivity. The Tampnet model take all of these concerns on board and offers cost effective services, delivered in time and at guaranteed bandwidths and service levels, where the customer does not have to engage in complex, time-consuming and expensive consultant heavy non-core business fiber installation projects.

Examples of these only over the last 15 months are new subsea fiber installation projects to connect the Gina Krog, Edvard Grieg, Ivar Aasen, Cygnus and Mariner platforms. These are projects that have been completed within the project schedule, within budget and with zero serious HSSE incidents.

Alternative Methods and Technologies for Connecting the Fiber

From an unrepeated subsea fiber cable system, there is a great deal of flexibility in the connection type to the topsides of an offshore asset. The method will always depend on a number of factors, such as the nature of the asset (fixed or floating, jacket or gravity base, TLP or semi-submersible, fixed or variable heading etc.), water depth and availability of conduits from the seabed to topsides (J-tubes or I-tubes typically). Tampnet has experience with most of the tie-in and pull-in options possible, and most of the platform asset types. Most recently, several tie-ins using wet-mateable subsea connectors on high power Raman pump links have been designed and installed. We used subsea wet-mate connectors to tie into a platform umbilical on a structure near a platform at one end of a 250km+ fiber segment. To add to the excitement, this whole tie-in design had to be placed only a few meters away from a large subsea isolation valve structure within the platform 500m safety zone, and be designed for a 30 year design life in a high current, high erosion environment. Tampnet managed the design work for creating the complete solution in cooperation with

our key installation and engineering partners, and the hardware has been serving the platform in question for about a year now.

Often, the pull-in to the platform becomes a significant cost factor in tying in to existing structures, and possibly even a prohibitive factor. Oil & Gas companies like to re-use conventions and techniques that are used for pulling in umbilicals and larger flowlines to platforms when designing fiber cable pull-ins, which quickly makes the cost escalate. Tampnet has been able to design very cost effective ways of engineering and executing both simple and complex platform tie-ins, lowering the overall project costs and maximizing the product quality and safety, while minimizing installation time and the impact to the platform space, operations and resources.

Anders Tysdal has worked as the CTO of Tampnet for 9 years. During his time in the company, it has grown from being a small telecom operator serving 34 fields in the North Sea to the largest low latency offshore telecommunications carrier in the world, serving more than 250 fixed and mobile offshore assets with high capacity and low latency communications. Tampnet now owns and operates a global network based on 2500km subsea fibre optic cables, about 100 offshore microwave line of sight links and approximately 70 offshore GSM and 4G LTE base stations. Tampnet is also a turnkey supplier for implementing both greenfield and brownfield offshore high capacity and low latency telecommunication solutions. Anders has led the work of developing technology and solutions for, and expanding the reach of, Tampnet networks both organically and through several M&A processes. He holds a Master's degree from the Norwegian University of Technology and Science and currently resides in Houston with his family.

2016 INDUSTRY REPORT
SUBMARINE CABLE MAP
2017 INDUSTRY CALENDAR

MORE INFORMATION >

SALES OPEN

CONTACT: Kristian Nielsen
+1 703.444.0845 | knielsen@subtelforum.com

BREAKING NEWS!

Nextgen Group Completes Australia's First Purpose-Built Subsea Fibre Optic Cable Connecting Darwin To Port Hedland.

By NextGen Press Release

Darwin September 14 2016 – Two years and 2100km later, Nextgen Group and Alcatel-Lucent Submarine Networks (ASN), part of Nokia, today officially "switched on" Australia's first purpose-built, subsea, fibre-optic network, servicing the oil and gas industries and the growing need for world-class telecommunications services in Australia's North West growth corridor.

The North West Cable System (NWCS) creates a link from Darwin in the Northern Territory to Port Hedland in Western Australia, connecting the INPEX-led Ichthys LNG Project offshore facilities and the Shell Prelude Floating Liquefied Natural Gas (FLNG) facility, 200km offshore in the Browse Basin, to their onshore data centres and business headquarters.

The state-of-the-art, high-speed data communications is now integrated into Nextgen's 17,000km national transmission network and Metronode's national network of data centres, providing immediate benefits to those communities in the Northern Territory and Western Australia connected to the cable.

Nextgen Group CEO, David Yuile, said the US$139m cable system not only provided essential support for Australia's offshore oil and gas projects but promoted competition with new telecommunication infrastructure for businesses and consumers in regional communities at the cable landing points.

"High-speed connectivity has now emerged as a vital component to operate any business regardless of geographical location – be it 200km offshore – or in remote and regional Australia.

"The completion of the NWCS is a significant milestone for the Nextgen Group and will deliver a digital dividend to

Australia's resources sector and to regional communities in Australia's North West," Mr Yuile said.

Philippe Piron, President of Alcatel-Lucent Submarine Networks said: "We are pleased to have supported the Nextgen Group with ASN's innovative technology to offer higher bandwidth availability, greater reliability and lower latency which are of critical importance to enhance the efficiency of operations on and off-shore.

"The NWCS project also demonstrates how technology advances can accompany the evolution of economic and energy needs, while enhancing the communication infrastructure in the region." Mr Piron added.

Nextgen has included additional capability in the system that allows for enhanced telecommunications resilience for the Northern Territory. It has also provisioned future connections into other locations such as the Tiwi Islands and established infrastructure-based, high-bandwidth capability in the Pilbara region.

The launch event in Darwin was attended by the Senator The Hon Matthew Canavan, Federal Minister for Resources and Northern Australia and the Northern Territory Minister for Corporate and Information Services, Lauren Moss, MLA.

About Nextgen Group

Nextgen Group is a leading provider of network connectivity and 10 Australian-based data centre facilities – in Canberra, Wollongong, Sydney, Perth, Brisbane, Melbourne and Adelaide – for Australian businesses, government agencies and telecommunications service providers. We deliver innovative and customised solutions to our customers, keeping them at the forefront of technological change. www.nextgengroup.com.au

INTEGRATION AND VISUALIZATION OF WELL FIBRE OPTIC DATA

BY DIGITAL ENERGY JOURNAL

Dynamic Graphics Inc. (DGI), a company specializing in data visualization and analysis, has developed a computer tool to display acoustic and temperature data, generated from fiber optic cables in wells, together with all of your other reservoir data and models.

This should make it easier for oil and gas companies to work with well fiber optic DAS and DTS data, to understand what the data is showing, and how it relates to what else is happening in the surrounding reservoir.

"My experience is that a lot of data is being captured but not used to its full capacity," said Jane Wheelwright, Technical Application Specialist at DGL, speaking at the Finding Petroleum forum in London on Apr 13, "Transforming Subsurface Interpretation".

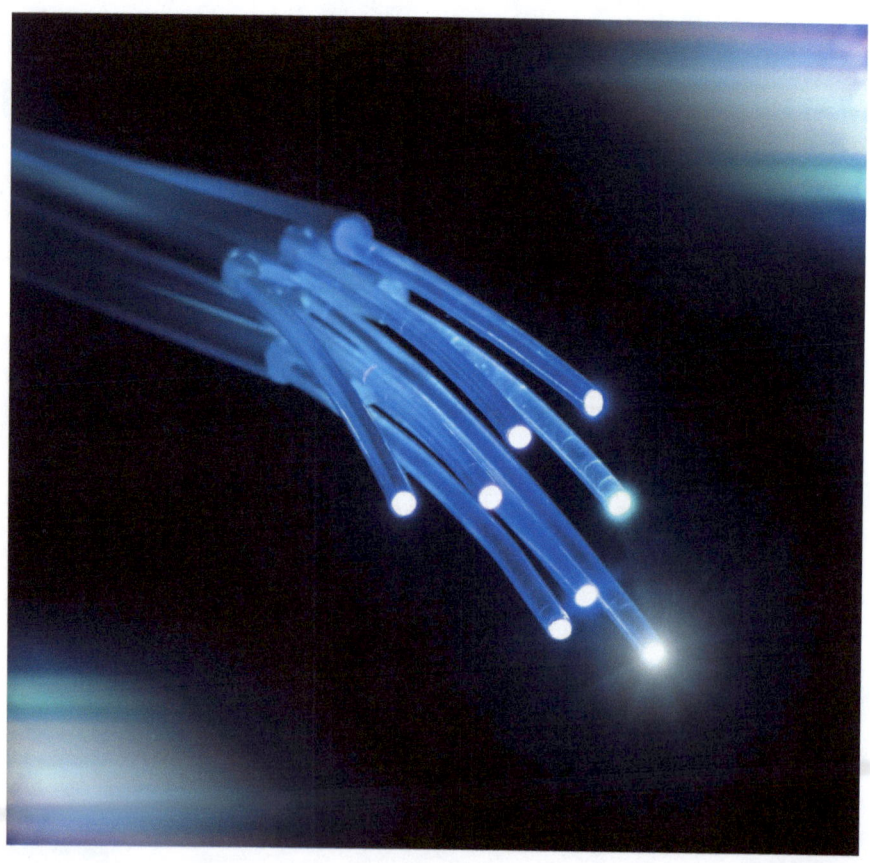

Also, "a key thing is not looking at a well in isolation but getting an understanding of what else is happening in the reservoir," she said. "The results are more powerful when they can be integrated with other data. It gives a far better understanding of what is happening with a well."

Displaying It

Dynamic Graphics provides two ways of displaying well fiber optic data. Firstly the temporal downhole view, where the temperature or acoustic data for the well can be visualized together with any other well and reservoir data, at any specific point in time.

This enables the data to be viewed at different points in time with other time varying data – so, for example, you could see a 4D seismic survey from a year ago, together with the acoustic data from the well at that time.

Another way is the 'waterfall' display where you see how the data recorded from the well is changing over time.

"The capability to see both these displays is very useful, they complement each other," she said.

"You want to have capability to integrate with existing data, quantitively and qualitatively, and be able to back interpret and cross plot the information.

CoViz 4D

Ms. Wheelwright showed a view of a mature offshore oilfield using Dynamic Graphics' CoViz 4D software, with many different data sets available.

By looking at the field at different points in time, you can see how the status of each well changed, and the history of the field. You can see the various fluid flows, with oil, gas and water in different colors.

"As water is injected and oil produced, changes can be seen," she said.

On the same view, you can see a 4D seismic data set, a geological model, and a reservoir simulation, drawn from other software pack- ages (in this case Seis-works, Landmark Nexus, and RMS).

The CoViz 4D software has a Global time slider which amalgamates all the time steps from each file, so you can see the correct status of all the data loaded at different points in time,(for example, hourly production data and yearly updates to the 4D seismic).

You can look in detail at any part of the field. For example, you might want to look at data from a producer and injector well pair.

With temperature data, you could see a thermal slug moving up the well in real time.

You can see well temperature data together with the well completion data, because elements of the completion could be causing a change in temperature.

If you have production data together, you could see (for example) that the temperature rose after the well was shut in.

You can see where valves were open and closed.

"There are wide ramifications in using this technology," she said. "It gives quite unique data and information for the reservoir management team, and it allows for understanding what is happening in a reservoir and optimizing decision making."

Well fiber optics

Fiber optic cables can measure acoustics and temperature by understanding how they change the flow of light through the cable. One cable can provide many different sorts of data.

DAS ("Distributed Acoustic Sensing") is a way to use fiber optic cables in wells to make an acoustic (sound) recording of what is happening in the well.

This can be used to spot problems (for example damage to a component), and also to record seismic data from inside the well.

DTS ("Distributed Temperature Sensing") is using the fiber optic cable to measure temperature at all points in the well.

Fiber optics are already used to replace production logging tools, as a means of under- standing which zones in a well are providing most of the oil, and to spot flow leaks inside the well.

The systems are also used to monitor for restrictions in the well bore (which can cause the fluid flowing past them to make a sound). It provides information about completions. It can be used in fracking, to monitor how well the fracking is going.

Oil companies need the data to work out the best way to frack a certain well, and the best way to optimize hydrocarbon recovery.

The data generated by DTS (temperature) is much smaller volume than the DAS (acoustic) data. Most DAS files are larger than 1 TB, which makes them very hard to work with. "A key issue is to reduce the size of data files," she said.

The DAS data is currently usually provided in HDF5 format, which can be imported into CoViz 4D.

INVESTORS DO NOT BELIEVE OUR STORY

BY DIGITAL ENERGY JOURNAL

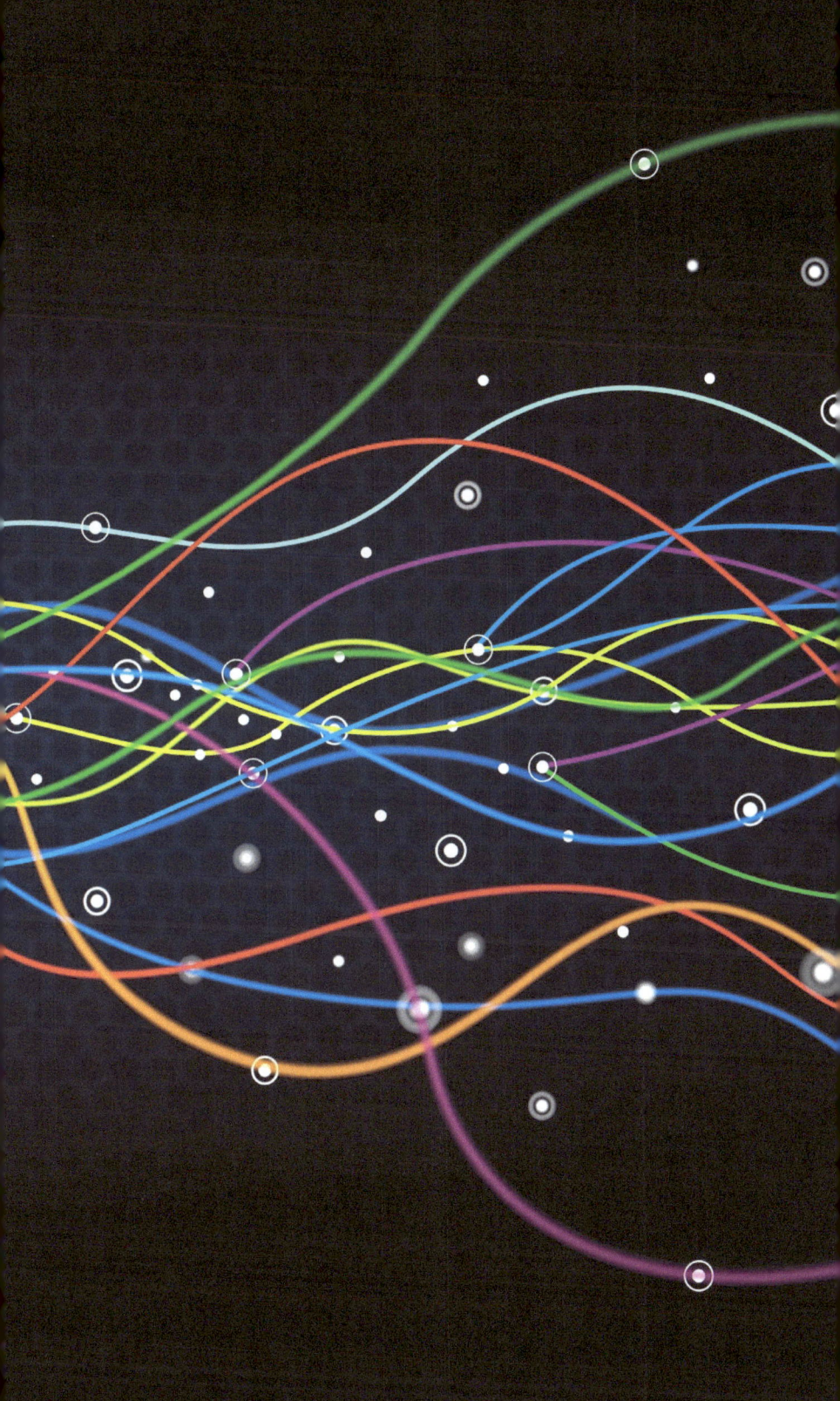

Investors are getting very skeptical about the oil and gas industry's ability to find oil and gas, says Finding Petroleum Director David Bamford. Can better use of technology lead to more exploration success - and make us more investable?

Many oil and gas investors are starting to lose faith in the oil and gas industry's ability to deliver with exploration, said David Bamford, director of Finding Petroleum and a non-executive director of Premier Oil, speaking at the Finding Petroleum event in Stavanger "Transforming Subsurface Interpretation".

"The viewpoint of all these investors is, we don't believe a word that you say."

"There's a big issue of actual failure to deliver exploration results and reserves additions, to the point at which investors don't believe our story. That's a serious problem."

This is in addition to concerns from low oil prices and high operating costs.

This means that the oil and gas industry does not look very attractive compared to other industries investors could put money into, he said. That leads to the question of whether there can be better ways to understand

the subsurface with the help of new technologies, to gather, integrate and manage data.

Mr. Bamford suggests seabed seismic recording, fiber optics in wells, gravity and gravimetry, as technologies worth looking harder at. You have to choose the right technology for your project, and then, perhaps hardest of all, integrate all the data together at the end.

Seabed seismic

Seabed seismic is about recording seismic data on the seabed, rather than with recording devices towed behind a vessel on the surface of the water. You can get a

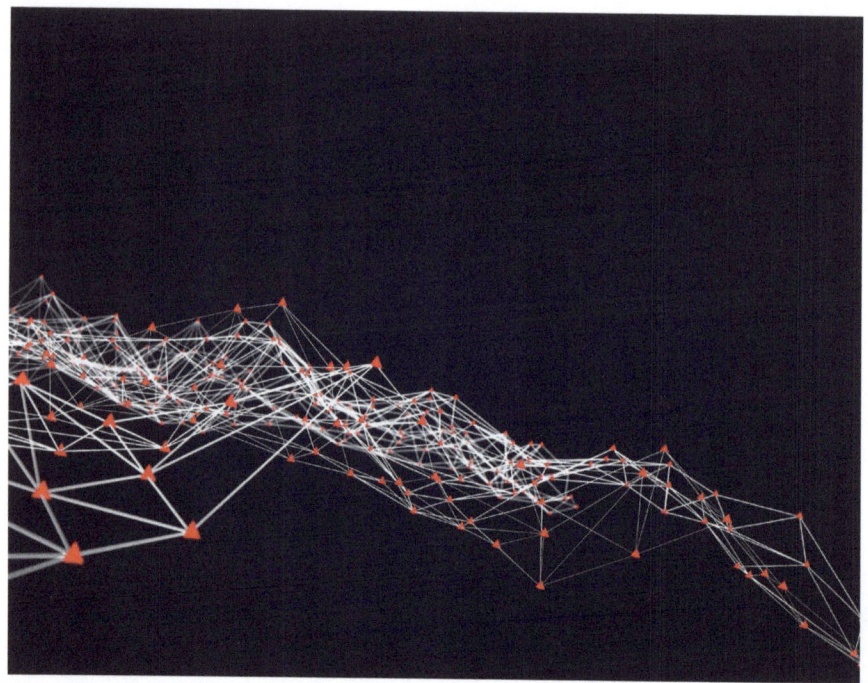

much higher resolution recording if you make it on the seabed.

"The idea that you would replace towed streamer seismic with seabed based acquisition has been around for quite a while in the form of [seabed] cables," he said.

"Universities in the UK and Norway have been using seabed acquisition for years and years and don't understand why our industry doesn't."

But "the operational difficulties has driven several companies out of business," Mr. Bamford said.

One seabed acquisition company "had such operational difficulties, it was having to repeat surveys at its own costs. It eventually disappeared in smoke."

Companies are moving towards "nodes" - individual devices placed temporarily on the seabed - which can prove cheaper to deploy than permanent cables.

With seabed recording, you can acquire four component data (compressional waves and three directions of shear waves). You can use this additional data to more sophisticated processing, including monitoring and mapping fractures.

"The whole thing improves the chance of prediction [of oil]," he said. "This is a technology which has arrived and is finding widespread use."

Almost all of the seabed surveys done so far have been done in the Gulf of Mexico, and have been "proprietary" surveys (where one oil company contracts the seismic contractor, rather than "multiclient" work, where the seismic contractor doing work which is then sold to many different companies). They are also mainly with nodes, rather than cables.

Fiber in wells

Meanwhile, the use of fiber optics in wells is making big steps forward, particularly the way that they can be used as listening devices, using technology mainly developed for the defense industry.

The cable is "so robust, usable and inexpensive you can deploy it in any well," he said.

"It doesn't interfere with production," he said. As you can imagine, production engineers really hate any technology in wells which interrupts flow.

"It delivers [continuous] recording of the noise the production is making, [plus] valves closing and shutting, anything going on in the well. So you can pinpoint where petroleum or water is flying into a well very accurately.

"You can use it for downhole seismic profiles [recording seismic data in the well] and record them any time you want. There is a possibility of permanent reservoir monitoring facility with some fiber optics downhole."

"So quite powerful stuff."
The technology has been used to record seismic data in the well by one Middle East client, to monitor a multiwall steam injection in enhanced oil recovery, and recording Vibroseis shot data.

"There are obvious applications, but now really operationally sensible, because this stuff is so robust and works so well," he said.

Gravity

Meanwhile gravity recording, or more specifically "Full Tensor Gravimetry" (FTG), is "very effective," he said. "Of all advances in technology, that one had the most impact on finding oil and gas."

With FTG the gravity is recorded by two devices a short distance apart on the same aeroplane or ship, and then you make a comparison between their recorded signals.

The same noise (for example from aircraft move-ments up and down) can be recorded by two sensors, and by putting their signals together, the noise can be cancelled out. This means you end up with a much better signal to noise ratio.

Mr. Bamford got familiar with FTG in his previous role as a non-executive director of Tullow Oil, which was using the technology in Uganda and Kenya.

"In Uganda, we shot about 10,000 km2 of this Full Tensor Gravimetry. You could integrate it pretty well with 2D seismic, and you'd have a real exploration database you could explore with."

"In Kenya we acquired 60,000km2 of this gravity data. In somewhere like Kenya you can acquire tens of thousands of square kilometers relatively inexpensively, such as $2-3m dollars for the 60,000km2."

"It really is good at demonstrating the basin shape and showing you the structural pattern."

Offshore, FTG is being used to help solve subsalt problems, which are proving hard to image using seismic only.

Another interesting non-seismic geophysics technology is Controlled Source Electromagnetics (CSEM). "Particularly in Norway, there's been a lot of talk of the help that CSEM can provide to the exploration mapping process," he said.

The challenge is working out how to fit EMGS data into your 3D seismic.

"If you look at the equations involved and the rock parameters involved, it is not obvious how you join these things together," he said.

Choosing a technology

So which technologies should you use where?

For example, in a survey in North West Africa, with complex carbonate rock, "the critical technology to deploy is going to be ocean bottom nodes," he said.

For North West Europe, typically with sandstones and shale, "a combination of nodes and electromagnetics might be helpful to you."

"There's been a reportedly quite large fractured basement discovery in the UK West of Shetland called Lancaster," he said. "Allegedly this is part of a new play that could open up all the way from Ireland to Norwegian Sea."
"The key is first of all mapping where the edge of the basin is and what's its history. [Here] Full Tensor Gravimetry (FTG) would be the important technology."

The value of the reservoir depends on the distribution of the fractures, so these would need to be understood. "You can envisage some combination of the technologies I talked about earlier helping you," he said.

Integration

The critical question which is not well answered is how to integrate all the data together.

The science and equations behind the technologies has been understood for over 100 years.

"But these sciences do not talk about the same thing. Some have first order differential equations, some have second order. That's a complicated answer."

Most geophysicists" IT set-ups are designed for interpreting 3D seismic.

"The work processes that are built around them don't easily allow the integration of other data," he said.

"If you accept that these things will change and transform what we do, and therefore increase success rates, reserves extensions, appraising discoveries, making new discoveries, then somehow we need to figure out how to integrate these things," he said.

wfnstrategies

celebrating **15** years of excellence

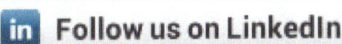

Telecoms consulting of submarine cable systems
for offshore Oil & Gas applications

JOHN PENDER:
THE CABLE KING

BY STEWART ASH

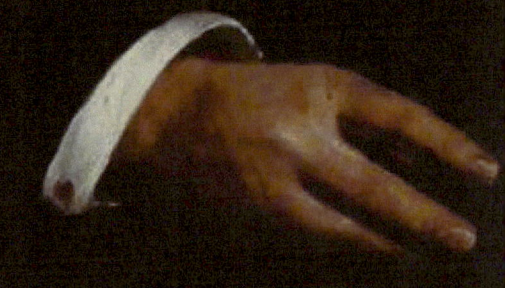

As many readers will know, this year is the 150th anniversary of the completion of the first successful telegraph cables to be laid across the Atlantic. The 1866 cable was completed between Valentia, Ireland and Heart's Content, Newfoundland, on 27th July. Just six weeks later the 1865 cable was finally completed on the 8th September 1866. Celebrations on both sides of the Atlantic have commemorated this historic event.

However, there is another anniversary that is arguably more significant for the submarine cable industry: 2016 is the bi-centennial anniversary of the birth of the man who probably did more than any other individual to make the Atlantic Telegraph a success. He then went on to found a submarine telegraph cable empire that encircled the earth and earned himself the epithet 'Cable King' before his death. That man was John Pender (1816-96).

John Pender was born on 10th September 1816 in the village of Bonhill in the Vale of Leven, just 24 miles to the northwest of Glasgow. He was the middle child of seven of James Pender and Marion née Mason.

Wee Field' Bonhill 1830

Burn Street Bonhill

In 1810, the Penders had moved from Campsie in Stirling to Bonhill so that James could take up a job in one of the printing and bleaching businesses that had grown up in the area. John Pender's obituary, published in the *Dumbarton Herald* in July 1896, indicates that James worked at the Bonhill Printworks known as *'Wee Field'*. The family lived in a cottage on Burn Street and from 1823 John attended the village school, also in Burn Street. He apparently showed a natural aptitude towards mathematics and drawing.

Sometime between 1824 and 1829, the Pender family moved to the Gorbals, which at that time was an area of up-market residences for the merchant classes, a mile or so outside the city of Glasgow to the south of the river Clyde. This was the boom period for the Gorbals, and moving there at that time suggests a significant rise in the Pender family fortunes. John was sent to Glasgow High School to continue his education. Unfortunately, all of the early nineteenth century records of the school were destroyed in a fire some years ago and no details of his academic performance have survived. John left school at the age of 14 and took up an apprenticeship as a *'Pattern Maker'* at Croftengea, one of the Bonhill calico print works.

In 1835, Croftengea became John Orr Ewing & Co, when it was taken over by John Orr Ewing (1809-78) and Robert Alexander, where they began producing 'Turkey Red' dyed products. John Orr Ewing was a business associate and friend of James Pender and, on completing his apprenticeship in 1837, Orr Ewing facilitated his son's advanced into a management position. A report of 1839 in the *New Statistical Account of Scotland* described the company as employing 192 men, 142 women and 104 children with an output of close to three million yards of printed goods per year.

John Pender married Marion Cairns, the daughter of a Glasgow tailor, on 20th November 1840. The parish records give his profession as *'Calico Printer'*. Marion quickly presented John with a son, James, born in Bonhill on 28th September 1841. However, she died just a few weeks later, on the 16th December, her twenty-second birthday. The cause of her death is unrecorded but it was most likely due to complications related to the birth of her son.

Croftengea Bonhill

The business of John Orr Ewing & Co thrived and expanded, selling their Turkey Red products in Glasgow and Manchester, the center of a growing export trade to China and India. John Orr Ewing was making a fortune! In late 1843 he decided to retire and sell his shares in the business to his partner, which he finally did in 1845.

The two John's would become lifelong friends and it was Orr Ewing's decision to quit the business, combined with the recent loss of his wife, that encouraged Pender to make a new start by moving to Manchester and start making money for himself. In January 1844, he set up his own business, John Pender & Co 'Commission Agents' with offices at 20 David Street and took up residence at Grove House in Higher Broughton, then a small township to the north of the city. The detached house was on the main Manchester road, and he lived there with his two-year-old son James and his youngest sister Marion who was his housekeeper.

Over the next few years John Pender & Co flourished and John moved his offices to 29 Dale Street and his residence to Bredbury Hall in Stockport. On 12 June 1851, he married Emma Denison (1816-90), an heiress from Daybrook in Nottingham, whose ancestry can be traced back to mid-sixteenth- century landed gentry. Emma encouraged John to diversify his investments and so when the English & Irish Magnetic Telegraph Co was launched in Liverpool on 10th June 1852, she suggested that he should take a major stake in the company. On 23rd May the following year, the chief engineer of the Magnetic, Charles Tilston Bright (1832-88), oversaw the installation of the first successful cable across the Irish Sea from Port Patrick to Donaghadee and a telegraph service between London and Dublin was then set up. Pender closely followed the development of this service and it was this that stimulated his lifelong interest in the electric telegraph.

Crumpsall Hall c.1850

Pender's textile business continued to grow and by 1856 he sold Bredbury Hall and moved to the larger estate of Crumpsall Hall on the Middleton Road to the northwest of Manchester. His family then comprised Emma and James plus Henry Denison Pender (*b.* 8th Oct 1852), Anne Denison Pender (*b.* 9th Nov 1853) and John Denison Pender (*b.* 10th Oct 1855).

In October 1856, John Watkins Brett (1805-63), Charles Tilston Bright and Cyrus W Field (1819-92) came to Liverpool and Manchester promoting the Atlantic Telegraph Company and Pender was one of the first to take shares in this company. Although he was appointed a director of the Atlantic Telegraph Co he did not take a leading role in the project at this stage. At the end of that year his last child, Marion Denison Pender (*b.* 4th Dec 1856), was born at Crumpsall Hall. The following year Pender was appointed Chairman of the British & Irish Magnetic Telegraph Co, which took over the English & Irish company and would provide a vital link in the Atlantic Telegraph.

John Pender c.1855

Pender appears to have had very little to do directly with the 1857 attempt and 1858 failed Atlantic Telegraph cable. When the joint British Government and Atlantic Telegraph Co investigation report was published in April 1861, Pender was focused on his textiles business, because the early part of the American Civil War (12th April 1861 – 9th May 1865) had created a cotton famine in Manchester and alternative sources had to be found.

On the 12th December 1862, John Pender was elected the Liberal MP for Totnes in a by-election and, in order to undertake his parliamentary responsibilities, he purchased a London residence at 18 Arlington Street.

It was Richard Atwood Glass (1820-73) who recognized that if an Atlantic Telegraph was to be successful it would need a single company responsible for all aspects of the project. Unfortunately, Glass did not have the standing or reputation to make this happen. However, in late 1863 he shared his thoughts with Cyrus Field and Field took the idea to Pender. Pender believed such a thing could be possible so he took on the task. He was reappointed as a director of the Atlantic Telegraph Co on 17th March 1864, and then oversaw a merger between the Gutta Percha Co and Glass, Elliot & Co. On 4th April 1864, the Telegraph Construction and Maintenance Co known as Telcon was formed with Pender as its first Chairman. To achieve this outcome Pender had put up a personal guarantee of £250K. One month later Telcon was awarded the contract for a new Atlantic Telegraph.

As part of Pender's grand plan, on 14th January 1864 a consortium led by Daniel Gooch (1816-89), and Thomas Brassey (1805-70) supported by John Pender purchased Isambard Kingdom Brunel's (1806-59) ship, the SS *Great Eastern*, at auction in Liverpool for £25,000. A new company, the *Great Eastern* Steamship Co, was established with Gooch as chairman and Pender and Brassey as directors. She was converted for cable work and, as part of the refit one of her five funnels (second from the stern) was removed to make way for a cable tank. She was then chartered to Telcon for £50,000 worth of Telcon's shares.

As is well known, the 1865 cable lay failed when the cable parted just 600 nautical miles short of Newfoundland. Daniel Gooch was on board, and on the return passage he wrote a letter to a friend expressing confidence that they would return the following year and complete the task. Prior to the *Great Eastern* sailing

from Sheerness on 15th July, John Pender had been on the husting where he had been re-elected as the member for Totnes at the General Election on 12th July.

To Gooch's disappointment, the Atlantic Cable Company was fully extended. New capital was required to keep the dream alive, and due to the American Civil War, none could be expected from America. Once again, it was Daniel Gooch and John Pender who answered the call. They raised £600K of new

The SS *Great Eastern* 1865, by Henry Clifford (1821-1905)

investment, co-founding the Anglo-American Telegraph Company in March 1866. Both became directors of this new company, and Richard Atwood Glass was appointed as its first Chairman. This company took over Field's New York, Newfoundland & London Telegraph Company, appointing Field as a non-executive director. Unsurprisingly, a contract to build a new Atlantic cable was given to Telcon; the fee for this was paid in Anglo-American shares.

The SS *Great Eastern* sailed from Sheerness on 30th June 1866, and to confound long-held superstitions, the lay from Valentia commenced on Friday the thirteenth, with the well-known outcome, described earlier.

A number of the key men involved in the Atlantic Telegraph were recognized by Queen Victoria for their contribution to the success of this massive undertaking. Captain James Anderson (1824-93), commander of the SS *Great Eastern*, Richard Attwood Glass, the managing director and Samuel Canning (1823-1908), the chief engineer of Telcon, were knighted. Daniel Gooch and Curtis Miranda Lampson (1806-85) were created baronets. Lampson was originally an American from New Haven, Vermont but had become a naturalised British citizen in 1849. He had joined the board of directors of the Atlantic Telegraph Company in 1856, becoming its vice-chairman then, in 1866, he became a director of the Anglo-American

Telegraph Company. Although the contribution of Cyrus Field was recognized and greatly appreciated, it was considered inappropriate to offer an American citizen an English honor.

Despite his pivotal role in the final success of the Atlantic Telegraph, John Pender received no recognition whatsoever. He was chairman of the British & Irish Magnetic Telegraph Co, founded and was Chairman of Telcon, a director of the Atlantic Telegraph Co, founder and director of the Anglo-American Telegraph Co and was also on the board of the *Great Eastern* Steamship Co. He had taken more financial risk and almost certainly done more than any other individual to ensure the success of this project but no rewards came his way.

Why this should have occurred has never been made public but it is almost certainly due to the Totnes General Election. Shortly after the election, a petition alleging corrupt prac-

tices was brought by John Earle Lloyd and Edmund Tucker. This led to a House of Commons Select Committee hearing under the chairmanship of Edward Pleydell Bouverie (1818-89). Evidence was heard from 16-23 March 1866 and the outcome was that John Pender's election was declared void, and in addition, he was found guilty of bribery by offering Robert Harris, a local blacksmith and Conservative agent, a position worth £300 per year, if he voted for him.

This type of vote buying was common practice in a number of so called *'Rot-*

John Pender c.1870

ten' or *'Pocket'* Boroughs at that time. Although Pender strenuously denied these accusations and Harris was exposed as a convicted perjurer, the political mood was for clamping down on such electoral practices. On 6th June 1866, Queen Victoria ordered a Royal Commission to look into electoral corruption at the Great Yarmouth, Lancaster, Reigate and Totnes elections. The commission finally reported in March 1867; the report was a precursor to Benjamin Disraeli's (1804-81) Reform Act of 1867, under which Totnes was disenfranchised in 1868. The Queen's honors for the Atlantic Telegraph were made public on the 15 November 1866, so it would have been impossible for Pender's key role to have been acknowledged by the Queen at that time.

In 1868, Pender stood down as chairman of Telcon and set about building the submarine cable empire that would become the Eastern & Associated Telegraph Companies. By 1870, England was connected to India and from June 1872, messages could be sent from London to Sydney over Pender's cables.

In January 1873, John Pender sold his Crumpsall Hall estate and the family moved to Arlington Street. The entire contents of the house were sold at auction, promoting speculation in the newspapers that Pender was a ruined man. The truth was that both Arlington Street and Minard Castle, his summer estate on the northwest bank of Loch Fyne in Argyll, were fully furnished and there was no room for the extra furniture. Pender sold the Minard Castle estate at the end of 1875 and on 16th May 1876 he took out a 21-year lease on Foots Cray Place, a Palladian mansion in Kent, owned by Coleraine Robert Vansittart (1833-86).

Over the next few years his contribution to subsea telegraphy was recognized by many countries around the world, but it wasn't until 1888 that it

Foots Cray Place

was finally acknowledged in Britain, when he was knighted Knight Commander of St Michael & St George (KCMG). This was later elevated to Grand Cross of St Michael & St George (GCMG) in 1892. Interestingly these awards were both granted while Robert Arthur Talbot Gascoyne-Cecil (1830-1903), 3rd Marquis of Salisbury was Prime Minister. Cecil's London residence was at 20 Arlington Street, next door to Pender!

Sir John Pender died at Foots Cray Place on 7 July 1896 and is buried in All Saints' churchyard, Foots Cray, alongside his second wife Emma (*d.* 8th July 1890)

Sir John Pender c.1890

and his son Henry (*d.*13th January 1881). There is strong circumstantial evidence in the family papers to suggest that he was soon to be created Baron, but he died before Queen Victoria could sign the warrant.

Unquestionably John Pender was a major, if not the greatest, contributor to the success of the Atlantic Telegraph. In building his submarine cable empire he did more for submarine cables than any other man and undoubtedly deserved the title '*Cable King*'. His cable empire became Cable & Wireless, a name that like John Pender has now been consigned to history after the takeover of Cable & Wireless Communications by Liberty Global plc.

Stewart Ash's career in the Submarine Cables industry spans more than 40 years, he has held senior management positions with STC Submarine Cables (now Alcatel-Lucent Submarine Networks), Cable & Wireless Marine and Global Marine Systems Limited. While with GMSL he was, for 5 years, Chairman of the UJ Consortium. Since 2005 he has been a consultant, working independently and an in association with leading industry consultants Pioneer Consulting, Red Penguin Associates, Walker Newman and WFN Strategies, providing commercial and technical support to clients in the Telecoms and Oil & Gas sectors.

The Power of Submarine Information Transmission

There's a new power under ocean uniting the world in a whole new way. With unparalleled development expertise and outstanding technology, Huawei Marine is revolutionizing trans-ocean communications with a new generation of repeaters and highly reliable submarine cable systems that offer greater transmission capacity, longer transmission distances and faster response to customer needs. Huawei Marine: connecting the world one ocean at a time.

HUAWEI MARINE
NETWORKS

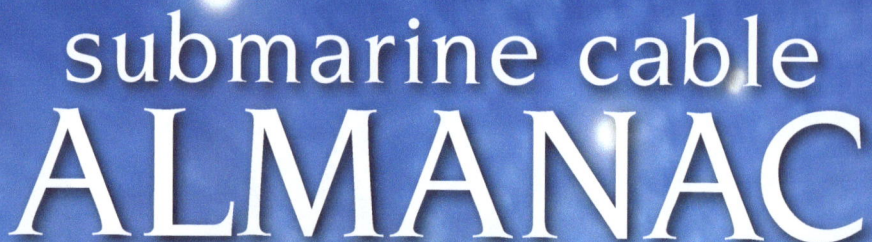

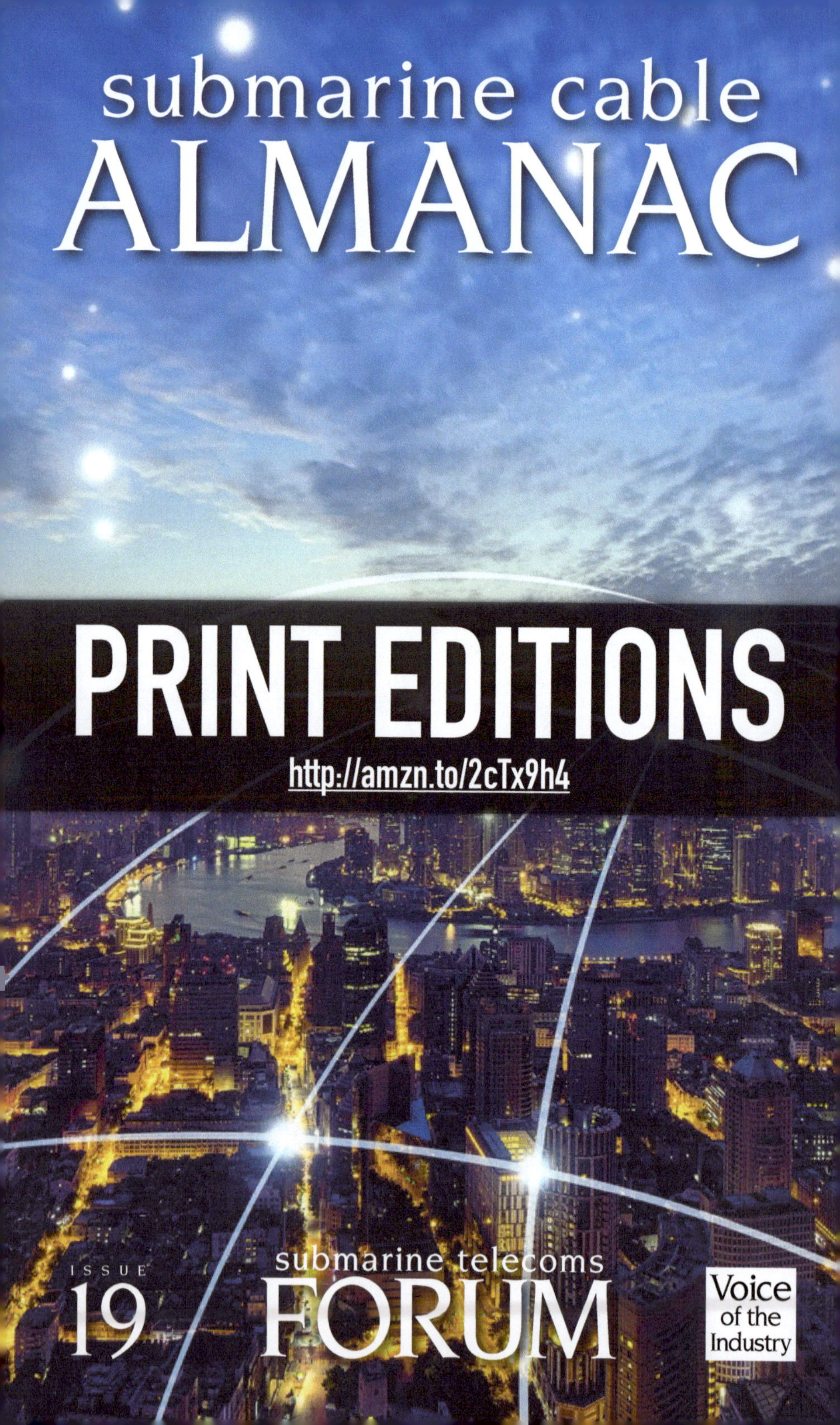
submarine cable
ALMANAC
PRINT EDITIONS
http://amzn.to/2cTx9h4
ISSUE
19
submarine telecoms
FORUM
Voice
of the
Industry

Sharks threats

Sharks have been periodically a hot topic for submarine cables. There was two main burst, the first one starting in in 1986, at the advent of optical fiber cables (and ten years after Spielberg's movie "Jaws"!). A second bounce was triggered in 2014. This interest was certainly amplified by the mythic interest for these amazing animals.

The first submarine cables provided early occasions to discover that the marine life could be encountered quite deep, with occasions such as the entanglement of a sperm whale in 1877 at a depth below 1000m off Basil. Fish bites were encountered in early times of telegraphic cables, but completely forgotten later, in particular at the early times of the optical era. Historical reports can be found in a chapter by Gérard Fouchard in Reference 0.

The first Suboptic event took place in Versailles in 1986 and was the first large marketing event of this kind dedicated to the submarine cable community. Submarcom, a joint venture of the French industry, asked a cartoon designer, Michel Haillard, to produce a short movie that showed how robust their submarine equipment was, in particular their bright new optical fiber cable. For this purpose, they imagined an improbable attack by a pink shark. You can see some snapshots from the movie in Figure 1. The shark picture was used to illustrate a virtual attack

and absolutely not at all a real case. The pink color is there to clarify this virtual nature. The movie was a permanent presentation on the Submarcom conference booth 24 hours a day.

At the end of the conference, there was a presentation of the AT&T optical

Figure 1:
Snapshots from the marketing Suboptic cartoon of Submarcom in 1986

repeatered system Optican-1 deployed in 1985 in Canarias by AT&T with Telefonica. Refer to the previous edition of SubtelForum magazine about another early optical system Antibes Port-Grimaud deployed in 1984. At the end of the presentation, the speakers from Telefonica and AT&T announced that they had suffered in one year two shunt faults (short circuit between the conductor and the sea) on

this cable. The investigations accused shark attacks, and the presenters showed as a proof a sample of the cable having a shark tooth deeply stuck into the cable (not easy to find due to the very same color as polyethylene). This announcement during Suboptic, amplified by the Submarcom cartoon, was the starting point of a major crisis in the business since sharks were perceived as endangering the deployment of the new optical technology. The funny detail is that, despite its cartoon, Submarcom was not at all aware of any possible problem with sharks. But nobody could believe it and they participated in the inflation of the story. The international press grabbed the topic with suggestive titles: "Shark bites halt trans-Atlantic cable", "Submarine cables: shark threats"

Fault continued on Optican-1, leading to four faults in total within several ten kilometers only. AT&T found not less than 50 shark teeth imbedded in their cable, in the vicinity of the faults. They showed that a guilty party was a rare deep-dwelling crocodile shark that is found in water depths below 1000 to 2000m and was biting the cable during its quiet period on the sea bottom, that was perceived as a real threat for the long term.

The real surprise was that the previous generation of coaxial cables did not suffer such problems. And the question was to understand how sharks could attack specifically cables of the new optical technology. Many new cables were in the planning phase and primarily the first transatlantic cable TAT-8, that encouraged very large scale investigations of shark behaviors, leaded first by AT&T Bell Labs involving the marine scientist community. They studied the species of sharks attacking Optican-1 by analyzing their teeth. Then numerous behavior trials were also conducted, both at sea and in controlled marine aquaria.

Sharks are quite sophisticated creature and can use powerful tools to detect their preys, that permitted them to survive since more than 400 million years! Sharks had been a topic of academic research, and the stimuli for biting were well studied: the factors are color, smell, movement, electro-magnetic fields, or simply the curiosity of the shark (taste testing). The electro-magnetic sensitivity of sharks became a subject of specific investigations for AT&T, since optical cable had a larger electrical current and induced more fields than the previous co-axial generation. But at the end, the experiments to identify the specific factors were not convincing, and no bite could be reproduced by the sole electric field. Movement inducing acoustic attraction were also known as an important factor from previous cases of attacks on buoy mooring, and acoustic stimuli induced scarce non reproducible experimental shark attacks during AT&T experiments.

The highlight of the investigations was then a 9 week campaign at sea CAN/MARE with worldwide scientists to analyze the statistics of shark species along the route of TAT-8. Two hundred and eleven (211) sharks were caught (Greenpeace and the Sea shepherd were not so active at that time…), but no "Crocodile shark" specie that had attacked Optican-1.

At the end, the studies with sharks were not really convincing. The understanding of Optican-1 was finally that the deep rock floor of Canarias induced some cable suspensions favoring shark attacks. AT&T focused then on the protection of cable. The armoring used in the shore was more than sufficient in these areas, but the deep sea cable is quite naked. AT&T developed a deep sea light armor compatible with laying and recovery below 1500m with an iron protection tube around the polyethylene. The so call "shark bit" cable protection became a customer requirement. The story is detailed in Reference 1.

The competitors had no such experience of shark bites. They were also seeking advice with scientists with moderate results. A researcher from the Centre d' Océanologie de Marseille even expressed that the case was "interesting" because it illustrated openly an unexpected fallout of pure research for an industrial application! As reported in Ref 0, a study in Japan highlighted that sharks could attack cables in case there is food around, or following mechanical vibration of the cable. At the end, to answer the market expectation, all suppliers had to develop a light weigh protected cable.

The main interest of these protections is that they were also able to protect from the abrasion on rocks that happen with cable suspensions on sharp rocks, even with very low currents. Could it be the main cause of shunt faults on optical cables ahead of shark bites? The term "shark bite" protection was coined. Many different terms for cable protected against abrasion were used after shark bite: Light weight protected (LWP), Fish bite protected (FBP), Special purpose application (SPA), and Light weight screened (LWS). See Reference 2.

TAT-8 studies were proceeding and in 1987, AT&T presented their shark threat studies to the consortium and opened the search of shark bites on older cables to all operators. The analysis was then leaded by the International Cable Protection Consortium (ICPC) who collected the available data. There was only one questionable case of shirk bites reported during the history of coaxial cables, but it was soon understood that these coaxial generations were bigger cables with a metallic wrapping, and were thus less sensitive to consequences of bites. Back to the previous telegraph cable generation, 36 cases of fish bites were recorded since 1900, and surprisingly, it provided also the opportunity to rediscover that 16 cases of whale entanglement happened during this time as they were not so rare as today. ICPC also stressed the cable protection implications, and the care needed to improve the cable routes and the laying, to minimize the cable suspensions favoring fish bites but overall abrasion that are detrimental to the long life of a cable. The state of the art nowadays is to armor the cable down to 1500m and to use light armoring from 1500m to 2500m depth when the sea floor is at risk concerning suspensions.

The topic was then closed, but bounced 30 years later: A short movie was put on Youtube in 2010 showing a shark wandering and biting once a cable before leaving with an obvious disappointment. You can still access the link below.

https://www.youtube.com/watch?v=1ex7uTQf-4bQ

Figure 2 :
Shark attack on a submarine cable

This movie taken during a survey had a limited success during several years, growing up to 2014, when it was hyped during the 2014 annual "Shark Week", a media event promoting sensational news involving sharks. At the same time, people had started to understand that their precious Internet relied on submarine cables. Could sharks endanger the Internet? The number of hits on the Youtube exceeded one million!

Very soon after the "Shark Week", Google had planned to communicate on the launching of their biggest investment in submarine cables in the Pacific Ocean, FASTER Cable Network. They did not ignore the web fever on the topic and they naturally announced that they protect the FASTER cable against sharks, but they added that it was by a "Kevlar" layer. It was noticed by some readers that cable protection existed already and did not use Kevlar. But this communication enhanced the crisis: "The Global Internet Is Being Attacked by Sharks, Google Confirms".

The experienced operators had to recall the established knowledge and to reassure their customers that there is no uncontrolled shark threat for their traffic. The ICPC took the initiative to refresh the database of cable repairs due to sharks (ref 3). The ICPC review recalled that from 1901 to 1957, a period of telegraphic cables, at least 28 cables damage were attributed to fishes. During 1959 to 2006, a period of coaxial cables and of early fiber-optic cables, around 11 cables needed repair due to fishes. This was only 0.5% of all cable faults! Since 2006, there was no declared cable fault attributed to sharks. This decrease with time was definitely assigned to the improved cable protection and better laying techniques with less cable suspensions.

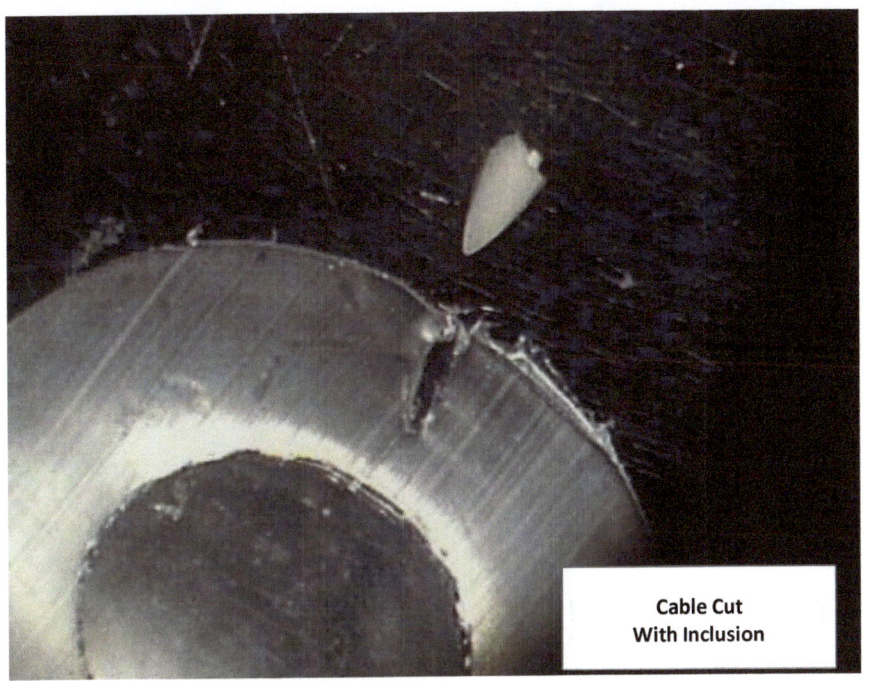

Cable Cut
With Inclusion

It still happens that sharks bite a cable close to the surface during laying, when the white "deep sea" unprotected cable moves behind the cable ship. The Figure 3 illustrates a case encountered close to Africa in 2010. But it is rare and happen only as an accident during laying, and not during the life of the cables.

Figure 3:
Photo of a cable cut with its shark tooth inclusion

The statistics are clear: the cable outages due to sharks are no longer relevant. A status of the root causes of cable cuts was presented in Suboptic 2016 (Ref 4). Figure 4 is taken from this presentation. One can recognize that fishing and anchoring are the main cause of cable cuts.

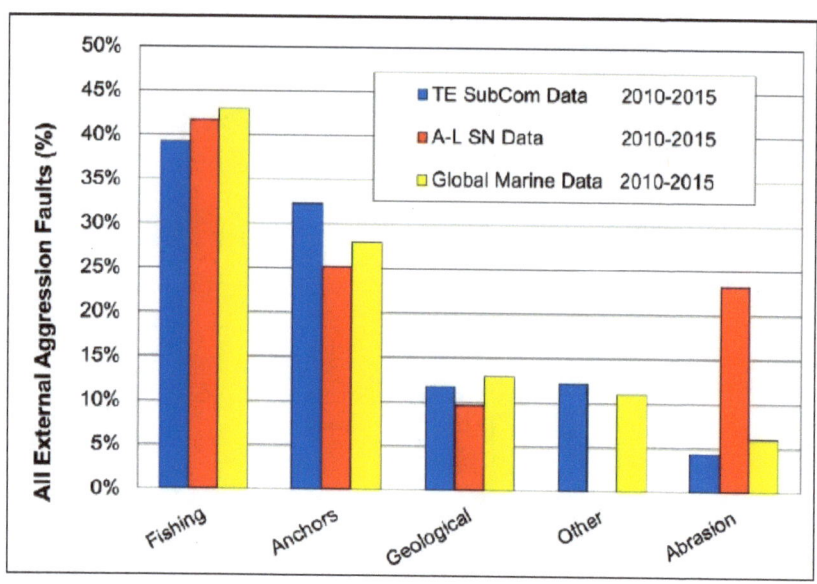

External Aggression Faults

Figure 4:
Causes of external aggressions on submarine cables

The public fear that sharks could endanger our internet has no relevant background. Cable cuts are finally rare and mainly due to external aggressions close to the coast. Fishermen are more dangerous for cables than fishes! One can be sure nevertheless that the rumor of shark threat will continue to bounce regularly.

References:

Ref 0: Du Morse à l'Internet, R.Salvador, G.Fouchard, Y.Rolland, A.P.Leclerc, Edition Association des Amis des Câbles Sous Marins, 2006 (book)

Ref 1: Marra, L.J., 1989. Shark bite on the SL submarine light wave cable system: History, causes and resolution. IEEE Journal Oceanic Engineering 14: 230–237

Ref 2 : Undersea Fiber Communication Systems, Ed.2,

José Chesnoy ed., Elsevier/ Academic Press ISBN: 978-0-12-804269-4 (book)

Ref 3 : Sharks are not the Nemesis of the Internet — ICPC Findings , press release, July 2015

Ref 4 : Maurice Kordahi et al, Global trends in submarine cable system faults, Suboptic 2016, Dubaï, 2016

José Chesnoy (jose.chesnoy@free.fr), PhD, is an independent expert in the field of submarine cable technology. After a first a 10 years academic career in the French CNRS, he joined Alcatel's research organization in 1989, leading the advent of amplified submarine cables in the company. After several positions in R&D and sales, he became CTO of Alcatel-Lucent Submarine Networks until the end of 2014.

He was member of several Suboptic Program Committees, and chaired the program committee for SubOptic 2004.

José Chesnoy is the editor of the reference book "Undersea Fiber Communication Systems" (Elsevier/ Academic Press) having a new revised edition just published end 2015.

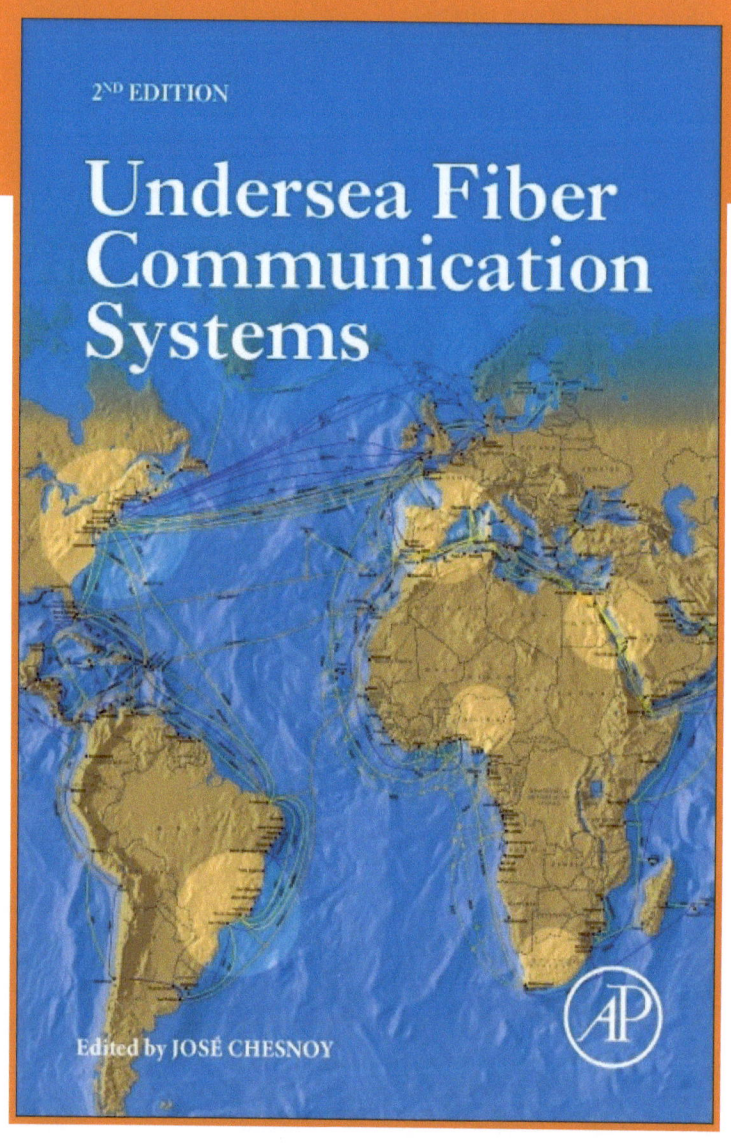

2ND EDITION

Undersea Fiber Communication Systems

Edited by JOSÉ CHESNOY

This comprehensive book provides both a high-level overview of submarine systems and the detailed specialist technical data for design, installation, repair, and all other aspects of this field.

Undersea Fiber Communication Systems, 2e
Edited by: José Chesnoy
Independent Submarine Telecom Expert, former CTO of Alcatel-Lucent Submarine Networks

With contributions of authors from key suppliers acting in the domain, such as Alcatel-Lucent, Ciena, NEC, TE-Subcom, Xtera, from consultant and operators such as Axiom, OSI, Orange, and from University and organization references such as TelecomParisTech, and Suboptic, treating the field in a broad, thorough and un-biased approach.

KEY FEATURES

- Features new content on:
    - Ultra-long haul submarine transmission technologies for telecommunications
    - Alternative submarine cable applications, such as scientific or oil and gas
- Addresses the development of high-speed networks for multiplying Internet and broadband services with:
    - Coherent optical technology for 100Gbit/s channels or above
    - Wet plant optical networking and configurability
- Provides a full overview of the evolution of the field conveys the strategic importance of large undersea projects with:
    - Technical and organizational life cycle of a submarine network
    - Upgrades of amplified submarine cables by coherent technology

ADVERTISER'S CORNER
BY KRISTIAN NIELSEN

Last year, right around now, I was planning a trip up to the Arctic for a project. I was excited, I'd never been up that far before, but I was mostly concerned with publishing the then looming Industry Report, a publication that's become expected. At the time, one of our resources had fallen suddenly and unexpectedly unavailable, and the future of the Report was fairly uncertain. In a stroke of luck, my travel was cut short and, in returning home, my analyst Kieran Clark had not only been able to fill the gaps, but had done so in more detail than we'd produced in years past! That edition of the Report was our most thorough and exhaustive released.

Now, a year later and with significant lead time, we're on the brink of releasing the latest edition of the Industry Report. Instead of being anxious, I'm thrilled to let you know that this edition raises the bar even further! I want to thank our research team and analysts for their tireless effort this summer – having read it, they've created an incred-

ible piece of research to share with all of you.

Of course, we're a few weeks away from release, advertising spaces are still available - if you'd like your ad to be one of those featured in this Report which will be downloaded a modest quarter million times, please feel free to drop me a line.

Ever yours,

Kristian Nielsen literally grew up in the business since his first 'romp' on a BTM cableship in Southampton at age 5. He has been with Submarine Telecoms Forum for a little over 6 years; he is the originator of many products, such as the Submarine Cable Map, STF Today Live Video Stream, and the STF Cable Database. In 2013, Kristian was appointed Vice President and is now responsible for the vision, sales, and over-all direction and sales of SubTel Forum.

 +1 703.444.0845

 knielsen@subtelforum.com

submarine telecoms INDUSTRY REPORT

MORE INFORMATION >

COMING IN OCTOBER

LAST CHANCE
10% DISCOUNT

CONTACT: Kristian Nielsen
+1 703.444.0845 | knielsen@subtelforum.com

Submarine Telecoms Forum, Inc.
21495 Ridgetop Circle, Suite 201
Sterling, Virginia 20166, USA
ISSN No. 1948-3031

PUBLISHER:
Wayne Nielsen
VICE PRESIDENT:
Kristian Nielsen
MANAGING EDITOR:
Kevin G. Summers

CONTRIBUTING AUTHORS:
Stewart Ash, José Chesnoy, Kieran
Clark, Digital Energy Journal,
Anders Tysdal

Contributions are welcomed. Please forward to the Managing Editor at editor@subtelforum.com.

submarine telecoms
FORUM

January:
Global Outlook

March:
Finance & Legal

May:
Subsea Capacity

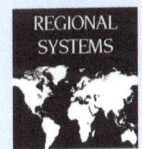
July:
Regional Systems

September:
Offshore Energy

November:
System Upgrades

Conferences

PTC 2017
15-18 January 2017
Honolulu, Hawaii USA
Website

SUBSCRIBE TO OUR FEED

JOIN OUR MAILING LIST

Voice of the Industry

CODA

BY KEVIN G. SUMMERS

It's haymaking time again, and, like clockwork, I've got an issue of SubTel Forum magazine coming out at the same time. But I'm not complaining, because while it's unpleasant to work all day in the field and all night on the computer, at least I don't have to sit in traffic every day.

When we made our first cutting of hay earlier this summer, I found myself questioning the life I had chosen. This was, literally, the second time I had ever been on a tractor, and we had absolutely no idea what we were doing. It was brutal, but we learned our lessons, learned our equipment, and even made some hay while the sun was shining. Good times.

Most of the problems we were experiencing the first time around were due to a missing piece on the tractor's lift arms. We're borrowing my neighbor's tractor, you see, and apparently the part in question broke once upon a time and he replaced it with something else. The replacement sort of worked, but my neighbor was mostly using the tractor for light mowing and not for the stressful haymakig that I was subjecting it to.

In other words, his makeshift solution didn't work under fire.

Having learned this lesson the hard way, I called up the local John Deere store and ordered a replacement part. I put it on the tractor, and amazingly enough,

it worked beautifully. I mean, what took me days of cursing and shaking my fist at the heavens last time around was just as easy as pie when we did it using the right part.

What does this have to do with submarine cables? You figure it out.

Kevin G. Summers is the Editor of Submarine Telecoms Forum and has been supporting the submarine fibre optic cable industry in various roles since 2007. Outside of the office, he is an author of fiction whose works include ISOLATION WARD 4, LEGENDARIUM and THE MAN WHO SHOT JOHN WILKES BOOTH.

 editor@subtelforum.com

Voice
of the
Industry